Solar Energy
A Do-It-Yourself Manual

Solar Energy

A Do-It-Yourself Manual

Charles Kiely

HAMLYN
London - New York - Sydney - Toronto

© Copyright Charles Kiely 1977
Published 1977 by
The Hamlyn Publishing Group Limited
London · New York · Sydney · Toronto
Astronaut House, Feltham, Middlesex, England
ISBN 0 600 38265 6
Produced and printed by The Compton Press Ltd.,
The Old Brewery, Tisbury, Wiltshire, England.

Second impression 1978

Cover Design, Robert Tilleard; Typography, Humphrey Stone;
Photographs, Strand Studios

Contents

Foreword

The practical use of solar energy is one of the frontiers open now to mankind. Moreover it can be used by ordinary people almost everywhere in the world. Through simple apparatuses built cheaply from locally available materials, we can harness the sun's free energy to heat water, cook, distill, produce electricity, and a host of other jobs. Solar energy is also harmless and cannot be turned into bombs. It cannot be cut off or diverted. It is clean and non-polluting. For these reasons it has attracted more attention than other alternative sources of energy.

I have written this book because I feel there is a need for constructional information about solar water heating systems. There is no magic related to solar panels; they are simple in operation and design, and relatively cheap to produce. They will save you money on fuel bills and make you a little independent of the established energy moguls.

CHARLES KIELY
Redditch 1977

1 The Application of Solar Energy

Energy crisis

Scientists and engineers have recognised the potential of the sun's abundant radiant energy for several hundred years. The first solar-powered heat pump was developed by a French engineer in 1615, and in the early nineteenth century several steam-driven solar engines were built by French engineers with the support of the government of Napoleon III. However, it was not until the early years of the twentieth century that solar energy equipment was developed at an appreciable rate. Many American companies supplying solar water-heating equipment in the 1920's enjoyed reasonable prosperity. Their development was spurred by a fast rising birth rate that created an energy gap in a sophisticated consumer society. The energy gap did not last long; soon the exploitation of fossil fuels was booming and the western world entered an era of cheap and abundant energy. Solar collectors became uneconomic when compared to cheap electricity, coal and gas.

Now that the age of cheap energy is at an end, we are beginning to wake up to the fact that the industrialised countries have been, and still are, squandering the earth's precious natural resources at an astronomic rate. We will soon be paying more for the privilege of using less energy and there will always be the danger of interruptions in supply if some governments use their valuable natural energy resources to 'influence' other less fortunate countries.

Fossil fuels becoming scarce

It is widely forecast that the earth's population will

double in the last quarter of the twentieth century. If this happens, energy demand will increase sharply and the crisis will worsen. If demand grows at the same pace as population growth then we may have completely exhausted fossil fuels within 30 to 50 years. Current forecasts suggest that supplies of oil will run out within the next 35 years, gas will last a little longer, and coal estimates vary from 50 to 200 years. It is a frightening thought and we must consider alternative sources of energy now if we do not wish to live in a cold, dark and inhospitable world.

Nuclear power, wave energy, geo-thermal energy, wind energy, water power, and methane gas are some of the alternatives being considered, and it is certain that they will all make some contribution to the future. However, we will concern ourselves with the inexhaustible power of the sun. After all, it pours down on us daily and costs nothing. It can be used to cook, heat water, produce electricity, distill water, dry crops; even to melt metal for industrial applications. It has been calculated that the solar heat radiated to the earth's surface in just one week equals the total present known fossil fuel reserves we are left with. The sun radiates the incredible amount of 170 million million kilowatts a year to earth; the amount of power the 3500 million people on this planet would use if they each left 48,000 1-kW electric fires burning continuously.

Not all of the sun's energy reaches the surface of the earth. Water vapour, dust, smoke, and our layer of atmosphere thin out the total radiance. But there is still enough, even in the unpredictable climate of the United Kingdom, to make substantial economies in bills for conventional fuels. Figures 1 and 2 show the amount of radiation falling on the British Isles during different months of the year. The totals are daily averages and are expressed in mega joules (to convert to kilowatts divide by 3.6). Southern parts of the British Isles receive approximately 1000 kWh (kilowatt hours) of solar energy per sq. metre per year. Northern England and Scotland receive 800 to 90 kWh.

Solar energy inexhaustible

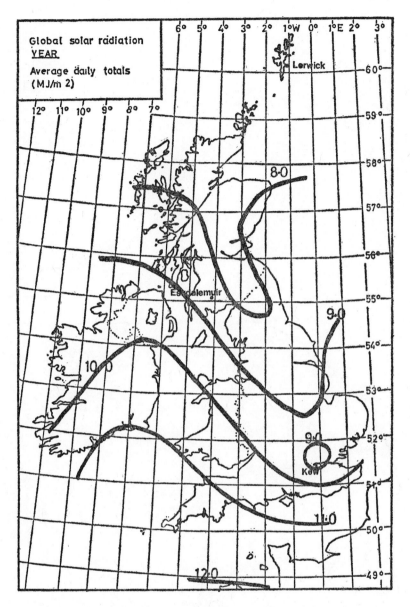

FIG. 1 *Annual mean daily solar radiation for the British Isles (mega joules per square metre).*

Reproduced from Solar Energy – a U.K. Assessment, © *1976 UK–ISES, courtesy of the International Solar Energy Society, U.K. Section, London.*

Annual variation of the mean daily totals of global solar radiation at Kew, Eskdalemuir and Lerwick

Data in MJ m⁻²

	Kew Latitude 51° 28	Eskdalemuir Latitude 55° 19	Lerwick Latitude 60° 08
January	1.94	1.49	0.81
February	3.59	3.94	2.66
March	7.38	6.26	5.61
April	11.48	10.48	11.75
May	15.87	13.97	14.45
June	17.35	16.32	16.68
July	15.81	13.75	13.99
August	13.34	11.32	10.95
September	10.17	8.05	7.32
October	5.65	4.29	3.24
November	2.49	2.01	1.14
December	1.47	1.15	0.56

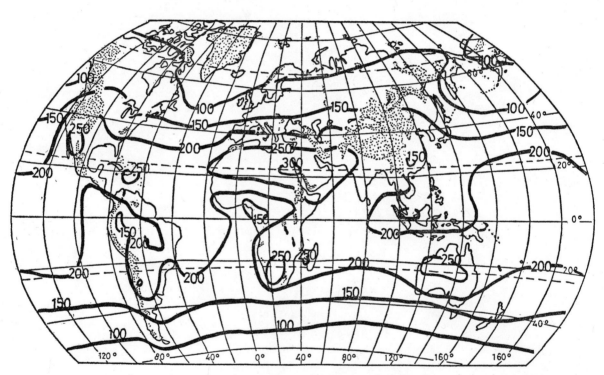

FIG. 2 *Annual mean global solar radiation on a horizontal plane at the surface of the earth (watts per square metre averaged over 24 hours).*

Reproduced from Solar Energy – a U.K. Assessment, © *1976 UK–ISES, courtesy of the International Solar Energy Society, U.K. Section, London.*

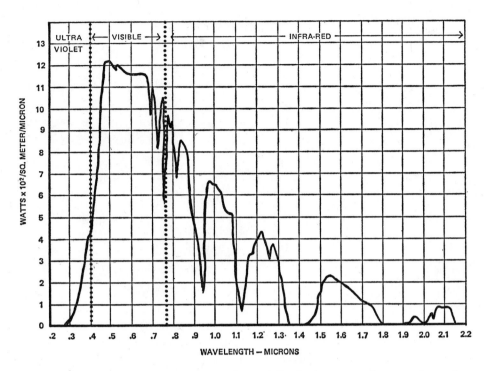

FIG. 3 *Standard solar radiation curve for engineering calculations.*

THE SUN

The sun is a star belonging to a numerous group called dwarf yellow stars. Nearly 80% of the sun's matter is hydrogen gas and most of the remaining matter is helium. The hydrogen atoms react by fusing together to create helium and this fusion creates the enormous energy that radiates continuously from the sun. Matter is turning into energy at the astonishing rate of $4\frac{1}{2}$ million tons per second giving a surface temperature of approximately 5500°C and an estimated temperature of 10 million°C at the core. The earth is travelling in a 600 million mile orbit around the sun. It is 91.33 million miles away in December and 94.48 million miles in July. It has been widely believed that the distance between the earth and sun varied at different times of the year because the earth's orbit was eliptical, but it is now believed that the earth's orbit is circular and the sun is slightly off centre. This means that the dead centre of the sun is not the dead centre of our universe.

The energy that radiates from the sun is electro magnetic radiation extending from tiny X-rays to radio waves 100 metres long with 99% of the radiation between 0.28 and 4.96M. There are three common designations of radiation – ultra-violet, visible and infra-red with most of the energy concentrated in the visible and infra-red groups. The sun's energy received on earth can be put into three categories, heliochemical, helioelectrical and heliothermal (helio coming from the Greek *helios* meaning sun).

Growth of vegetation

Heliochemical: This energy is the most important because it allows vegetation to grow by a process called photosynthesis. The green chlorophyll in plant life uses the sun's heliochemical energy to combine carbon dioxide with water to form sugar (for conversion to starch and cellulose) and oxygen. We are therefore continuously provided with energy (in the form of food, wood for burning, and fossil fuels when the plants have petrified) and oxygen to sustain life.

Solar cells to produce electricity

Helioelectrical: In recent years, notably the last fifteen, methods have been developed whereby specially treated wafer type cells can receive radiation and in turn produce electricity. This branch of solar utilisation is perhaps the most exciting, and although the cells are expensive, it is hoped that their cost can be reduced as production increases. More than 10% of the total energy received from the sun could be put to use as direct electrical energy. At present it would cost roughly £7,500 to purchase enough solar cells to power the average colour television set, but ten years ago it would have cost nearer £75,000, so you can see prices are definitely coming down! It is currently forecast that by the mid 1980's cells would be produced for a few pennies per watt. It is also worth noting that silicon, the element most cells are based on is the second most abundant element on earth.

Heliothermal: This is the energy we are most concerned with in building a solar water heating system, and it is the most widely used energy in the field of solar energy utilisation. When solar radiation falls on an object

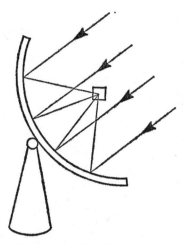

FIG. 4 *Cross section of typical solar concentrating (parabolic) collector unit.*

the temperature of that object rises until its heat loss equals its heat gain. Certain objects will lose heat more quickly than others because of their different material composition. The surrounding air temperature will also affect the rate of heat loss. Black bodies attract heat while polished metallic surfaces reflect it. The flat plate collector we are going to build is a flat black plate which will attract and store heat energy. Other types of solar collectors reflect the sun's energy, and because of their parabolic construction, focus the heat into a central point thus concentrating the energy (see Fig. 4). Flat plate collectors do not reach very high temperatures but they do collect most of the energy spectrum including some ultra-violet and all infra-red radiation and are suitable for use in northern climates. On the other hand, concentrating collectors can reach very high temperatures. At Odeillo in the French Pyrenees there is a solar furnace with a 30 sq. ft. collector and a huge parabolic mirror. It can reach temperatures of approximately 3000°C and is used for industrial processes, usually melting metals.

The efficiency of the flat plate collectors is increased by using a glass cover to create a greenhouse effect. Glass is transparent to 98% of the sun's radiation which concentrates its energy in the visible and near infra-red groups. When radiation has warmed up the collector plate it con-

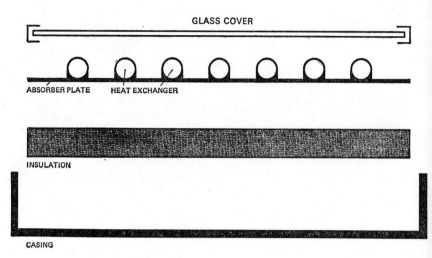

FIG. 5 *Exploded cross section through a typical flat plate collector unit.*

centrates itself in the far infra-red groups to which the glass is opaque. The glass then acts as a one-way valve trapping the sun's energy inside the panel. Temperatures of up to 204°C (400°F) can be reached in flat plate collectors and it is important to use construction materials that can withstand such high temperatures for prolonged periods (Fig. 5).

AVAILABILITY OF ENERGY FROM A FLAT PLATE COLLECTOR

The amount of energy obtained from a flat plate collector depends on the amount of energy it receives times the overall operating effciency. Most solar flat plate collectors (solar panels as opposed to solar cells) have an operating efficiency of between 40% and 60%, usually around 50%. To find the efficiency of a panel the amount of energy it receives must be accurately measured and compared to the heat input of the fluid to be heated, the flow rate, the temperature of the fluid output, and the surrounding air temperature of the panel. The panel described in this book will be approximately 50% efficient if constructed correctly; i.e. it will give up half the energy it collects.

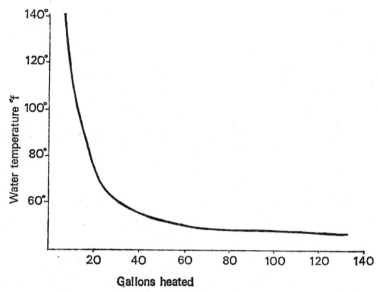

FIG. 6 *Average hot water output on a warm summer's day.*

Let us assume that you will be constructing three panels for your installation at the recommended size of 5′ × 3′. Each panel will have a collector area of just under 1.4 sq. metres (15 sq. ft.) giving a total area therefore of just under 4.2 sq. metres (45 sq. ft.). Assuming that the total annual radiation available in your area is 1000 kWh per sq. metre per year then the panels will collect

$$1000 \times 4.2 = 4200 \text{ kWh per year}$$
divide this by 50% (the efficiency rating)
$$4200 \div 2 = 2100 \text{ kWh}$$

So 2100 kWh is approximately the total useful energy that a three panel system will yield per year. Multiply this number by the local cost of one unit (kWh) of electricity and you will find how much money your solar water heating system is going to save. Normally, the amount saved per year will pay back the cost of construction and installing the system within a 3 to 4 year period. The system will carry on supplying approximately the same amount of energy each year for the rest of its working life; twenty years or more.

In the next chapter you will be shown how to construct a solar flat plate collector that will provide the energy to heat your domestic water supply. Chapter 3 will describe the various ways in which the panels can be installed and connected to your existing system. In subsequent chapters you will be shown how to use the panels for swimming pool heating, and how to construct a differential temperature controller to provide automatic control of the pump that circulates the water through the panels.

2 Solar Panel Construction

Tools and materials

The flat plate collector, or solar panel, is relatively easy to construct and should present few problems to the average handyman. If you can install a sink unit, plumb in a washing machine, make a bookcase, then this project should not be too difficult. An entirely novice plumber may need a little help. The panel can be constructed without the aid of special tools, specialised materials, or 'secret' processes. A hacksaw, wood saw, drill, hammer, screwdriver, nails, screws, paintbrush, file and a tape-measure are the tools you will need most. The materials needed are, 10mm ($\frac{3}{8}$″) annealed copper tube, 16swg aluminium sheet, timber for making a frame, glass fibre wool, marine-ply sheet, and 4mm (32 oz.) glass. These items can usually be obtained locally from good hardware shops, plumbers' merchants, or from mail order companies specialising in the D.I.Y. market. The actual sizes, quantities and specifications will be dealt with a little later on.

Efficiency

The panel has been designed with three main considerations. The first is efficiency. It is not the most efficient panel design but it does give very good results. Solar panels usually have an efficiency rating somewhere between 40% and 60%. This panel should give up approximately 50% of the energy it collects, making it comparable with most commercially available units. The second consideration is cost. You can construct this panel

Cost

for about one third of the price of a purchased panel, and even less if you can build from salvaged materials. The sun's energy is free, so the less you spend on installing the solar water heating system the sooner your initial outlay will be paid back, and the longer will be the period of free energy collection. Which brings us to the third consideration; durability. The longer the panel can provide useful energy the more attractive is the case for installing a solar water heating system. If carefully constructed, the panel should have a useful life expectancy of twenty years or more. The materials used will not deteriorate rapidly and as there are no moving parts there is nothing to wear out. There is virtually no maintenance required. Experience shows that dust and grime on the glass cover do not impede the collection of solar energy. Occasional checks to make sure that the glass cover is not cracked, the wooden frame is not warping, and the system has not developed leaks are all that is needed.

Durability

The panel uses copper tubing to carry the domestic water through its collector area. Copper is universally acceptable for plumbing and water systems. It does not corrode in oxygenated water, and so long as a suitable inhibitor is used it will not corrode in water/ethylene glycol mixtures either. Copper has high thermal conductivity properties, essential to provide efficient transfer of collected solar heat to the water supply. The existing plumbing in your home will almost certainly be copper and the new solar heating system should be copper also to avoid the corrosion problem associated with mixed metal systems. Copper is not cheap. However, the 10mm ($\frac{3}{8}$") copper tubing is mass produced and is readily available in 20 metre, 25 metre, and 30 metre coils. It can be purchased at most D.I.Y. central heating shops, or from your local plumbers' merchant. It is soft and easy to work into the desired shape.

Tubing

The copper tubing will be attached to the blackened absorber plate. This plate will collect the sun's radiant energy and transfer it to the water filled copper tubing. The plate should be made from aluminium or copper to achieve the best results. Aluminium, being cheaper than copper and more readily available in most areas, is a good proposition. A thickness of 16 swg (.064 inches) is preferable as thinner gauges tend to 'warp' under intense heat

Absorber plate

conditions and thicker gauges give reduced efficiency. The absorber plate will be painted with black paint and primer, and so will the copper tube, so there will be little risk of galvanic corrosion.

The absorber plate will be housed in a wooden tray. The tray is made from a hardwood frame with a marine ply backing sheet. A good preservative coating should be applied to the tray to ensure minimum maintenance during its exposed life.

Transparent cover

The panel should be covered with a transparent sheet to prevent heat loss in cool weather. Glass is preferred because it is readily available and has high light transmission value. It will 'trap' long wave radiation thus improving the efficiency of the panel. 4mm (32 oz.) float glass should be used. Acrylic sheeting can be used, and this will result in a lighter panel. Acrylic is usually much cheaper than float glass, but bear in mind that glass does not scratch or discolour. The transparent cover can be placed in the tray and sealed with wood beading and a suitable gum or resin.

Other items you will need are: glass fibre wool (the type used for loft insulation), a few sheets of aluminium foil or highly reflective 'lunar foil' or Mylar Aluminised Film, black paint and primer and a collection of wood screws, nails, wire, etc. Plumbing materials will be discussed in the next chapter.

The measurements given in this book should be followed as closely as possible. The author has experimented with most types of solar collectors before deciding on the design given here, and recommends that there be as much standardisation as possible in the construction of the panels so that results can be easily compared. However, should you have access to less expensive materials of slightly different dimensions and properties, then by all means use them.

CONSTRUCTION DETAILS

By following these construction details you will be able to build a solar panel 5′ × 3′ giving just under 15 sq. ft. collection area. Two or three panels will be sufficient for the average household (four people) and you should refer to the following chart in order to ascertain how many panels you will require. This chart applies only to climatic conditions similar to those in the British Isles.

Domestic Heating Installations

No. of Panels required	Cylinder or Tank capacity
1	15 gallons
2	30–35
3	35–50
4	50–70

The panel must be able to drain down in cold conditions to prevent the water in the copper tubing from freezing. *Water in solar panels can sometimes freeze when the outside air temperature is slightly higher than freezing point.* The alternative is to use an indirect or closed circuit system and to add anti-freeze to the water. This requires a heat exchanger in the cylinder or tank so that the anti-freeze/water mixture is kept separate from the domestic water supply. The panels are versatile and can be used in either system.

Construction can be broken down into five main parts.

1. panel tray and frame
2. absorber plate
3. copper tubing
4. insulation
5. glazing

PANEL TRAY AND FRAME

The panel frame should be $5' \times 3' \times 3''$. Decide on the method for making the corners (Fig. 7) and purchase the wood accordingly. A good hardwood should be used such as teak or cedar. Most timber merchants will cut the wood to length, and machine it if desired. $3'' \times 2''$ section is advised. (Fig. 8 for section detail). Good corner joints are essential to achieve good weather-proofing. The rebates or lips for receiving the collector plate and glass can be cut out with a circular saw and the cut out strips used as beading for the glazing. After the frame has been con-

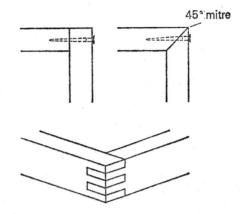

FIG. 7 *Corner Joints.*

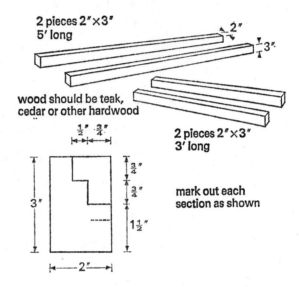

2 pieces 2″×3″ 5′ long

2″

3″

wood should be teak, cedar or other hardwood

$\frac{1}{2}$″ $\frac{3}{4}$″

2 pieces 2″×3″ 3′ long

mark out each section as shown

$\frac{3}{4}$″

$\frac{3}{4}$″

3″

1$\frac{1}{2}$″

2″

FIG. 8

structed two holes of 11mm ($\frac{7}{16}$″) diameter should be drilled in the sides of the frame to accommodate the copper tube inlet and outlet. They should be drilled 2″ from the top or bottom of the frame as required. If using two panels in your system then you will require one left hand and one right hand panel (Fig. 9) so that both panels connect at the top centre position. Inlet and outlet will be at opposite bottom corners to facilitate draining the panels in cold weather. If using three or more panels in parallel connection (Fig. 10) then all the panels can be of the same hand.

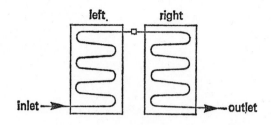

FIG. 9 *Two panels in series.*

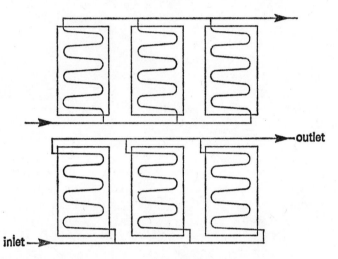

FIG. 10 *Three panels in parallel connection.*

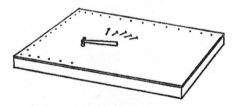

FIG. 11 *Fixing backing sheet.*

The frame can now receive its backing sheet. This should be of marine ply quality for best results, but 4mm ($\frac{5}{16}''$) exterior grade plywood can be used. Measure the frame exactly and cut the backing sheet accordingly. Place the sheet in position and nail to bottom side of frame (Fig. 11). An alternative method is to machine a rebate in the frame roughly $\frac{1}{4}''$ from the bottom and slide in the backing sheet before putting the fourth side of the frame in place. This will give a neater and more durable finish.

The whole frame and backing sheet should now be coated with a good wood preservative to ensure a lasting finish.

ABSORBER PLATE

Purchase and treatment of aluminium sheet

Purchase enough aluminium sheets of about 4′ 8″ × 2′ 8″ × 16swg for your needs (or an appropriate size to fit into the constructed tray). Look in your Yellow Pages for aluminium stockists and contact a few to gather quotations before deciding to purchase. Prices from stockists can vary considerably. Aluminium type H9 or similar is suitable.

Clean the aluminium on both sides and remove any trace of oil that may be present. Turn up the shiny side of the sheet and use some emery paper to etch the surface. This will remove small traces of oil and allow the paint to adhere to the surface better. Go all over the aluminium sheet in small circular rubbing movements. Next apply a good primer such as zinc chromate or red oxide. It is best to obtain the paint and primer in aerosol form to achieve the flattest possible coating on the absorber plate. It will also dry quicker and give an attractive finish. Small aerosol cans can be expensive for the amount of paint supplied, a large aerosol can will give better value for money. U-Spray comes in 400g (14oz) cans and is marketed by Borden (UK) Ltd., Southampton, SO5 9ZB. Their matt black is very good and has quite a high radiation absorption rate. Two cans will be enough to cover three absorber plates. The paint dries quickly (usually within ten minutes) and becomes very hard if left overnight (Fig. 12).

COPPER TUBING

Purchase and shaping of copper tube

The next step is to produce the serpentined copper waterway tubing. Using the diagrams (Fig. 15) work out how much copper tubing you will need. 30′ to 35′ (10 meters) is about right. The more copper tube the better, up to a point. The type to obtain is 10mm annealed. It is used extensively in central heating systems and is obtain-

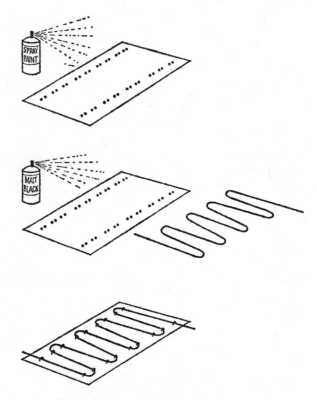

FIG. 12 *Apply primer to aluminium sheet. Paint sheet and tube.*

able from any good plumbers' merchant, D.I.Y. central heating shop, or from the numerous mail order central heating companies that advertise in magazines like *Exchange & Mart*. It comes in soft coils of varying lengths and is very easy to work into shape. If care is taken the copper tube can be shaped by bending it around a cylindrical object like a bottle (Fig. 13). A small hand bending machine such as the Preston Hand Bender is ideal and will ensure a perfect bend, resulting in good water flow through the system. As copper is a very soft metal it is easy to flatten or squash the tube when making a bend without the aid of a forming machine, so if you are bending by hand make sure that you bend the tube gradually. It may be time-consuming but you will avoid kinking the pipe which could result in furring caused by restricted water flow, thus reducing efficiency. Should you kink or squash the pipe you may be able to get it back into shape

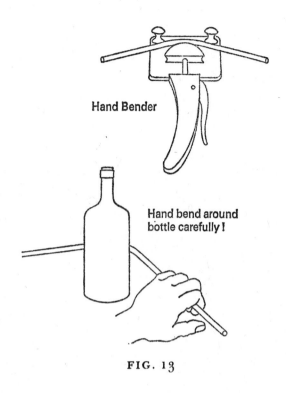

Hand Bender

Hand bend around
bottle carefully !

FIG. 13

by placing it on a piece of wood and tapping it with a wooden mallet. You will find the best method of bending the tube is first to draw off some tube from the coil and straighten it out on the floor (Fig. 14) so that you have a long straight piece of copper to bend rather than bending from the coil. Then keep bending (Fig. 15) until you have the required shape (refer back to Fig. 12). The formed tube should be hung from a piece of wire and sprayed with the matt black paint.

Attaching copper tube to absorber plate

When the paint is dry, place the formed copper tube on top of the aluminium absorber plate (black side) with a slight rise on each cross piece so that the water will always run down to the bottom outlet to drain (Fig. 16). Using white chalk mark out the pattern of the copper tubing onto the black absorber plate, and remove the copper tube. Next drill a series of small holes, two on either side of the cross pieces of tubing (Fig. 12). These holes will take the wire that will hold the tube closely against the absorber plate to achieve efficient heat transfer to the water in the tube. Next place the formed copper tube onto the plate and start threading a good strong wire through

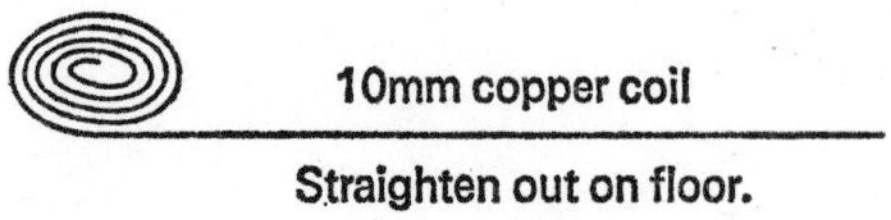

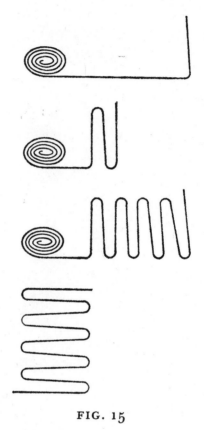

10mm copper coil

Straighten out on floor.

FIG. 14

FIG. 15

the holes and over the copper tube. A single piece of wire will do the job. Thread one end of wire through a starting hole, over the tube, and out through the second hole, continuing this lacing process until the final hole is reached and all of the cross pieces have been covered. The more loops the better to give good thermal contact between the copper tube and the absorber plate. An alternative method is to use small copper clips at each fixing point; they can be fixed in position with small self-tapping screws or small nuts and bolts. You may have difficulty in obtaining these clips as most small bore pipe clips are now

WRONG

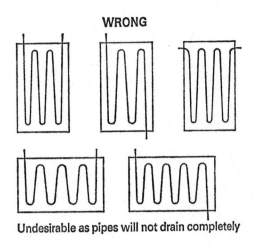

Undesirable as pipes will not drain completely

RIGHT

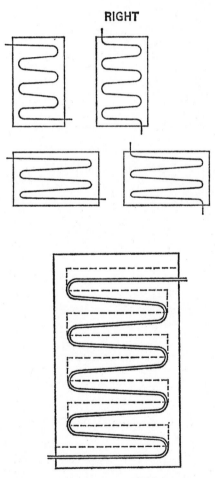

These pipe layouts will assist draining

FIG. 16 *Tube arrangement.*

plastic and unsuitable for high temperature work. Copper tube can also be soldered to the aluminium sheet using special solder and flux but the process has to be delicately controlled and is not for the amateur handyman.

INSULATION

Glass fibre wool and aluminium foil

The absorber plate is now ready for insertion into the panel tray, but first the insulation must be in place. 3" glass fibre wool should be laid in the tray. This will compress when the absorber plate is lowered on top of it (Fig. 17). A layer of aluminium baking foil can be sandwiched between the glass fibre wool and the plate with the shiny side to the plate. This will help stop heat loss by reflecting heat back to the plate.

Lower the absorber plate into the tray making sure that the inlet and outlet pipe ends are first bent back and then inserted through the 11mm ($\frac{7}{16}$") holes in the sides of the panel (Fig. 18). The plate can be held in place by strips of wood beading screwed to the sides of the inner frame. A final spray of matt black paint is needed to mask any scratches on the paintwork caused by the wire and to cover the inside wall of the wooden frame.

GLAZING

Inserting glass sheet

The panel is now ready for glazing. 4mm (32 oz.) float glass is the recommended quality for this size panel. 3mm float glass can be used in panels up to 10 sq. ft. but is too fragile for larger areas. The light transmittancy of 4mm glass is approximately 88% to 90% with a reflectance of 8%. Thicker glass will have slightly lower transmittance. Handle the glass very carefully and ask the glazier to deliver if this is possible. Measure the inside of the frame accurately and have the glass cut to size. There should be a gap of 15mm ($\frac{1}{2}$") between the top of the copper piping and the cover (Fig. 19). A larger gap will result in lower efficiency. After placing the glass in the tray, secure with wood beading and seal with a good sealant to keep out moisture.

Now that the panels are constructed they should be

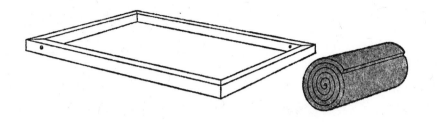

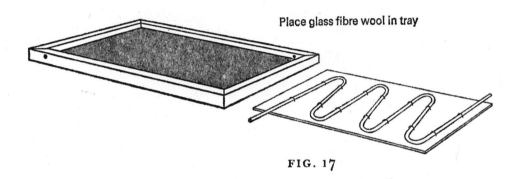

Place glass fibre wool in tray

FIG. 17

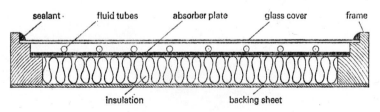

Insert collector plate

FIG. 18

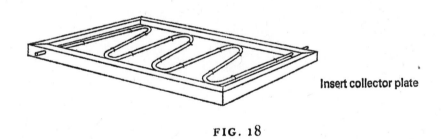

FIG. 19 *Cross section of solar panel.*

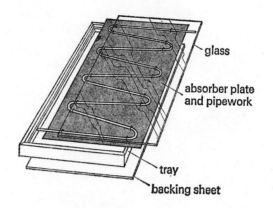

glass

absorber plate
and pipework

tray

backing sheet

FIG. 20 *Exploded diagram of a panel.*

stored in a safe place until required. Store them upright
so that the tube inlets and outlets will not be damaged.
The panels will be heavy and help should be obtained
when moving them.

3 Installing a Solar Water Heating System

Water needs to be circulated through the panels and stored in a suitable tank or cylinder for use when required. There are several methods of installation and we will give attention to as many as possible in this chapter. No great skill is needed but please read this chapter through carefully, perhaps several times, before beginning work. For those of you who are not familiar with plumbing joints there is an illustration of these included. We recommend that you use compression fittings throughout the system. They are more expensive than capillary but have the advantage of being simple to fit with a spanner, and they can be undone quickly if necessary. Wiring the pump and controller is fairly simple; follow the manufacturer's instructions and don't forget to use a low amp fuse in the controller input feed (see p. 70). Last of all, take your time! A well-installed system will last twenty years or more with very little maintenance, so care taken now will pay dividends later. Be careful when climbing and working on roof surfaces, for yourself and the people below!

Calculating numbers of panels required

The first task is to decide on the number of panels required. Between 1 sq. ft. and $1\frac{1}{2}$ sq. ft. of panel area should be used for every gallon to be heated. Two panels giving a total of 30 sq. ft. is normally sufficient area for the average house with a 30 gallon hot water cylinder. But of course some households use more hot water than others so the panel area should be increased accordingly. First determine the size of your hot water cylinder and work

Capillary joint (pre-soldered)

1. Clean pipe and inside fitting with wire wool.
2. Smear flux evenly on the inside of fitting and the outside of pipe ends.
3. Slide the pipe ends into the joint so each end stops against the shoulder inside the fitting.
4. Heat fitting with a blow-lamp until an unbroken ring of solder appears at each end of the joint.

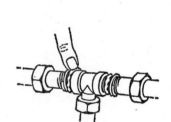

Compression joint

1. Clean pipe and compression rings with steel wool.
2. Insert the pipe into the fitting as shown in sketch, and smear a little jointing compound on the fitting.

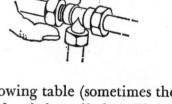

3. Bed the ring into the jointing compound, and screw the nuts on by hand.
4. When the nuts are hand tight, finish with a spanner, taking care not to over-tighten.

out the capacity from the following table (sometimes the capacity is stamped on the side of the cylinder). If you have more than four people living in the house then it is wise to multiply the capacity by $1\frac{1}{2}$ when calculating the number of panels.

It is best to over-estimate when determining the number of panels. Additional panels can be added at a later date but it is much more sensible to complete the installation in one go.

The next step is to determine the type of system to be used. There are two main types, thermo-syphon and forced circulation. Thermo-syphon is the simplest system to install because it needs no circulating pump or electronic controller. Water in the panels is heated by the sun and rises by natural convection to the storage tank or cylinder where it forces cold water out and back down to

Thermo-syphon system

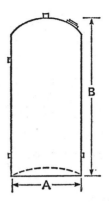

FIG. 21 *Domestic direct hot water cylinder to BS699.*

Size	A mm	B mm	A ins.	B ins.	Litres	Imp. Gallons
1	350	900	14	36	74	16¼
2	400	900	16	36	96	21¼
3	400	1050	16	42	115	25¼
4	450	675	18	27	86	19
5	450	750	18	30	98	21½
6	450	825	18	33	109	24
7	450	900	18	36	120	26½
8	450	1050	18	42	144	31¾
9	450	1200	18	48	166	36½
10	500	1200	20	48	200	44
11	500	1500	20	58	255	56
12	600	1200	24	48	290	64
13	600	1500	24	58	370	81½
14	600	1800	24	72	450	99

the panel to be heated (Figs. 32, 33 & 34). By leaving a minimum distance of 2ft. between the top of the panels and the bottom of the tank or cylnder, reverse flow is prevented at night. A simple non-return valve can also be plumbed into the flow pipe and will do the same job. Thermo-syphon is ideal in areas where there is no auxiliary power or where supply is unreliable. It works best in warm climates and is used extensively in Australia, Israel and the southern states of America. In northern latitudes it does not work so well as a good deal of heat is needed to get the water circulating rapidly. Also it rarely blends with the architecture of northern areas, as the panels must always stand below the storage vessel. Protection against freezing is another problem with a thermo-syphon system.

Forced circulation system

Forced circulation systems incorporate a pump to circulate the water. Panels can be placed above or below the storage tank or cylinder and can be situated some distance away if this is necessary. The pump is usually activated by a time switch or differential temperature controller to take full advantage of heat gain in the panels. This system is recommended for the majority of installations, because its greater efficiency usually compensates for a more expensive initial installation.

Direct system

Both systems can be divided into direct and indirect systems. With a direct system the water in the cylinder or storage tank is circulated through the panels and heated before use. An indirect system has a sealed fluid (usually water and anti-freeze) flowing through the panels. The fluid does not mix with the domestic water but flows through a heat exchanger in the cylinder which heats the water. A direct system is favoured because it is simpler, cheaper to install and more efficient. But the panels must be mounted above the cold storage tank so they can drain down in cold weather. If the panels have to be situated

Indirect system

below the storage tank then an indirect system should be used and a suitable anti-freeze mixed with the water in the panels. In a direct system the panels will drain automatically when the pump cuts out. With an indirect system it is recommended that an extra panel be installed to compensate for the drop in efficiency caused by using a heat exchanger. You will need at least $1\frac{1}{2}$ times the panel capacity that you would need with a direct system.

So to sum up. First determine the number of panels required, next decide on thermo-syphon or forced circulation, and third decide whether to use a direct or an indirect system.

Positioning panels

Panels must face towards the equator. In northern regions this means south, and in southern regions they must face north. This would appear obvious but the author *has* come across installations where the reverse has

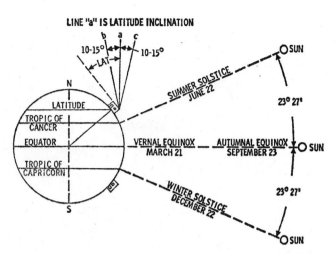

FIG. 22 *Position of the sun in the sky at noon and its relation to the inclination and location (latitude) of the solar collector.*

Reproduced from How to design and build a solar swimming pool heater, *1975, courtesy of the Copper Development Association Inc., New York.*

been done and the poor installer was left wondering why there was no hot water flowing from the panels. Small deviations can be tolerated, usually up to 15° east or west, but larger deviations will result in a big decrease in efficiency. The panels must also slope at roughly the angle of latitude to obtain maximum benefit all year round. A shallower angle of inclination will give better results in the summer months when the sun is higher in the sky, but in the winter when the sun is at a low angle very little energy will be collected unless the panel angle is increased. Figure 22 shows the sun's position at various times of the year.

Mounting panels

If you have a south-facing sloping roof of roughly the angle of latitude then you have no problems. If you do not have a south facing roof then you will probably need to build a frame and mount the panels on a south facing wall. The panels can also be mounted on outbuildings or even on the ground (Fig. 23). Do not place them too far away from the storage vessel as heat losses in the pipework will reduce the overall efficiency of the system. Which brings us to the next important point – LAG ALL PIPEWORK, and lag it well! Pipework inside and outside the building

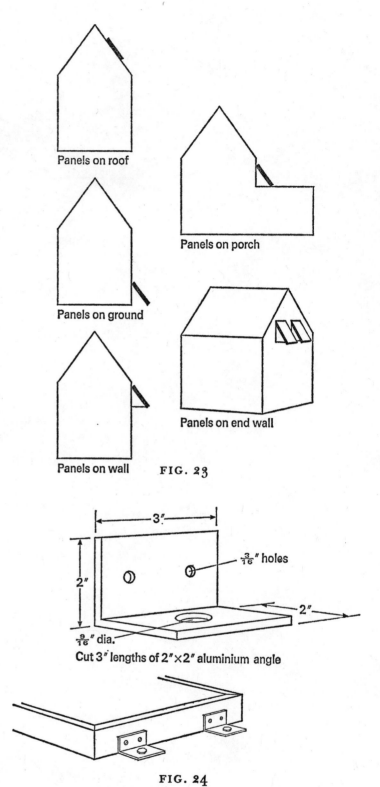

Panels on roof

Panels on porch

Panels on ground

Panels on end wall

Panels on wall

FIG. 23

$\frac{3}{16}$" holes

3"

2"

2"

$\frac{9}{16}$" dia.

Cut 3" lengths of 2"×2" aluminium angle

FIG. 24

must be covered with a good insulating material and the thicker the better. Foam lagging sections are very effective and cheap to purchase. They can be bound with tape for even better insulation.

The tools required for the average installation are as follows:

Power drill and extension lead
8mm ($\frac{5}{16}$″), 5mm ($\frac{3}{16}$″) and 14mm ($\frac{9}{16}$″) drill bits
Length of rope
Ladder(s) capable of reaching roof
Roof Ladder(s)
Set of open ended spanners or adjustable spanner
15mm ($\frac{1}{2}$″) pipe bending spring
No. 8 × 1″ screws (8 per panel) cadmium plated
Hacksaw
Screwdrivers

If you do not possess ladders they can usually be hired cheaply from a local builders merchant, or from a D.I.Y. store. Roof ladders are strongly recommended because they provide the necessary stability to ensure a safe and competent job.

Brackets

In addition to the above tools you will need plumbing fittings, but we will come to them a little later. You will, however, need brackets for fixing the panels to the roof. These can be made by simply sawing 3″ lengths from a strip of 2″ × 2″ × $\frac{1}{4}$″ aluminium angle. Four brackets per panel will be required. The angle can be obtained from any local aluminium stockist (see Yellow Pages) and is usually sold by weight. If you cannot obtain $\frac{1}{4}$″ (6.350mm) then use $\frac{3}{16}$″ (4.762mm). Three holes should be drilled as described; two at $\frac{3}{16}$″ and one at $\frac{9}{16}$″, or 5mm and 14mm (Fig. 24). The smaller holes will take the screws to the panels; the larger hole will take the bolt for securing the bracket to the roof. It is a good idea to fix the brackets to the panels before lifting the panels onto the roof as this means less work on the roof.

The bolts for securing the panels should be 13mm × 150mm long ($\frac{1}{2}$″ × 6″) and should be cadmium plated. Two cadmium plated nuts and washers per bolt will be needed. If bolts are unavailable, then ordinary steel threading (threaded rods) can be used, but make sure that they are plated! Ordinary steel nuts and bolts will react unfavourably with the aluminium brackets and rust very quickly. The plating prevents this from happening.

FIXING THE PANELS ON THE ROOF

First secure the ladders against the side of the building. Position the roof ladders so that the hook ends secure themselves as shown in Fig. 25. Sandbags can be used to support the ladders on the ground. Make sure that people are warned not to stand directly below the working area! You won't be thanked if you drop the electric drill on someone's head!

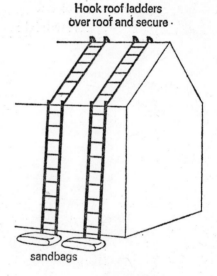

FIG. 25

The panels can be hauled up as shown in Fig. 26, or carried up by two people (Fig. 27). Be sure the rope is tied around the panel at all times so that the panel can be temporarily secured when it reaches the roof. Nerves of steel are sometimes needed when performing this part of the installation. If you have any doubts about your ability to erect the panels on the roof, then DON'T. Get a local builder to do it for you. It won't cost too much and at least you will have peace of mind (and no broken bones).

When the panel is in place, as near the apex as possible,

Two solar panels on a slate roof.

Ten panels mounted on an aluminium frame and linked in parallel. These panels are used for heating an outdoor swimming pool (not shown) in Warwickshire.

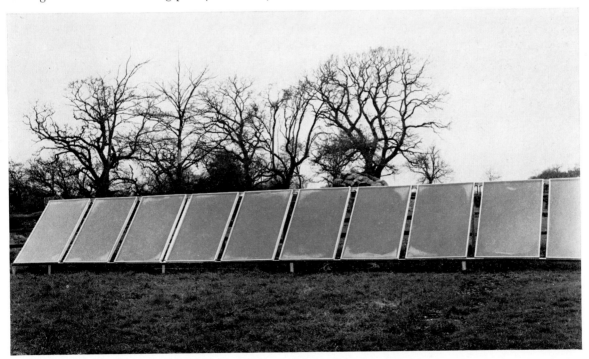

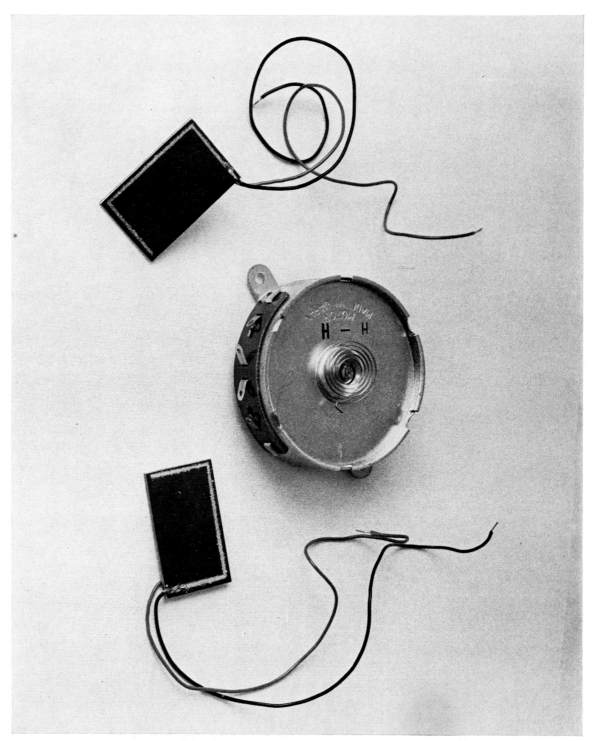

Two small selenium cells will power a low inertia electric motor in conditions of bright sunshine. These cells are only 1% efficient.

Three small silicon wafer cells power a small fan. Silicon cells are about 14% efficient.

The author heating water on a small parabolic (concentrating) cooker. Boils water in about 15 minutes with bright sunshine.

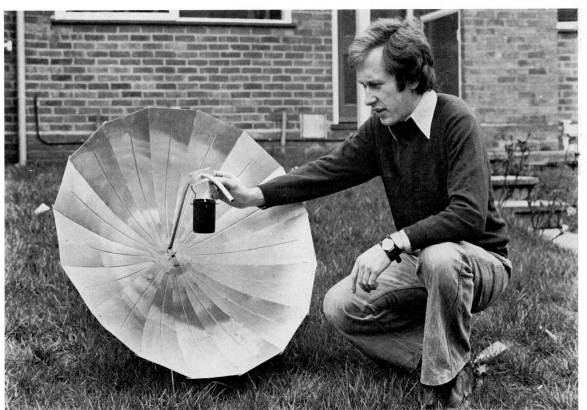

Bending the 10mm copper tube with a hand bender. A jig is held in a vice and the bend executed with a former (in the right hand).

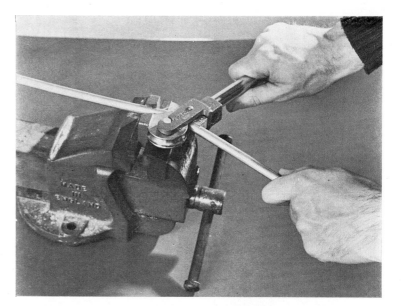

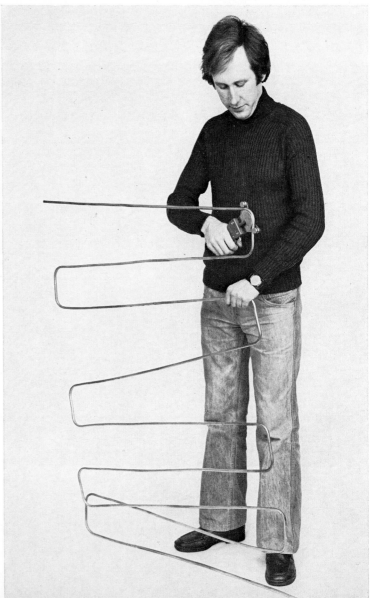

Bending the 10mm copper tube with a Rothenburger bending tool.

44

FIG. 26   FIG. 27

drill through the roof using the $\frac{9}{16}''$ holes in the brackets as a guide. Do not use a lot of pressure on the drill when drilling tiles or they may crack and split and use a slow-speed drill to reduce the risk of damage to the roof. Drop the bolts through the brackets and roof holes. They can be secured inside by using a small wooden batten as shown in Fig. 28. Two nuts should be tightened up against the wooden batten. All bolts and nuts and washers

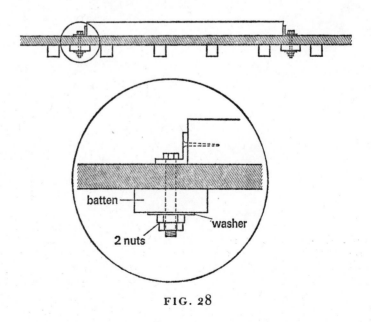

batten

washer

2 nuts

FIG. 28

Two panel installation

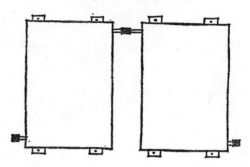

Three panel installation

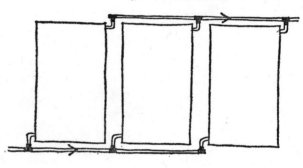

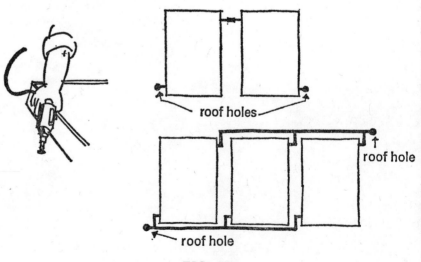

roof holes

roof hole

roof hole

FIG. 29

46

must be cadmium plated to avoid corrosion problems. Connect panels as shown in Fig. 29. Then drill holes to take the pipework through the roof. Connect 10mm × 15mm ($\frac{3}{8}'' \times \frac{1}{2}''$) reducers to the inlet and outlet pipes on the panels and connect to 15mm ($\frac{1}{2}''$) copper pipe and bring the pipe through the roof utilising a 90° elbow joint. Seal all the holes in the roof with a good waterproof sealant such as Bostick weatherproof sealant. (If you are

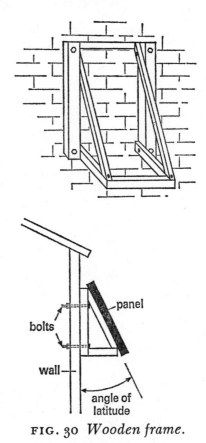

FIG. 30 *Wooden frame.*

using a differential controller in the system a small hole is required to bring the sensor wire through the roof.

If you secure the panels to a wall you will need a frame to mount them on. Use $3'' \times 3''$ or $3'' \times 4''$ hardwood or aluminium angle, $2'' \times 2'' \times \frac{1}{4}''$ or $2'' \times 2'' \times \frac{3}{16}''$. Bolt the framework to the wall with 13mm ($\frac{1}{2}''$) bolts and secure inside with at least two nuts and a large washer (Figs. 30, 31).

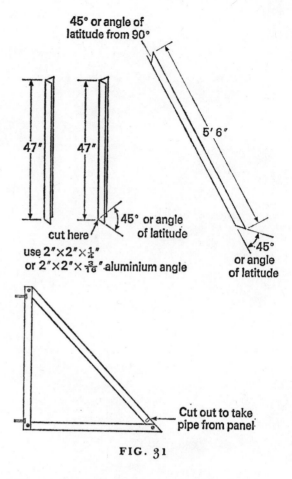

FIG. 31

Weather proofing

Note: Treat wooden frames with oil, varnish, or a good preservative for maximum protection. Aluminium frames will not need to be treated unless you live near the coast, in which case a good protective paint suitable for aluminium should be used. 'Alumin 100' paint is ideal. Coastal air will pit aluminium over a period of time and apart from looking unsightly, it may loose its strength after a number of years.

INSTALLING A THERMO-SYPHON SYSTEM

The panels can be connected to the existing hot water cylinder or to an additional pre-heat cylinder. Fig. 32 shows a simple one-panel system connected to a tank. Fig. 33 shows a two-panel installation and Fig. 34 shows a thermo-syphon system using a heat exchanger with a

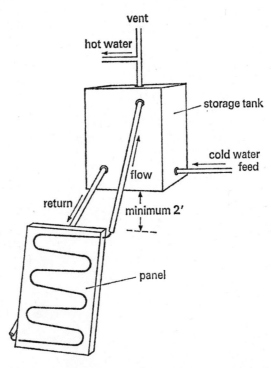

vent

hot water

storage tank

cold water
feed

flow

return

minimum 2'

panel

FIG. 32 *Typical thermo-syphon system using one panel.*

small expansion tank incorporated. *All indirect systems must incorporate an expansion tank* from which they may be filled and occasionally topped up with anti-freeze. N.B. Ethylene glycol is the correct anti-freeze to use in all indirect systems. It gets cold but does not freeze and has to reach 123°C before it will boil. A suitable inhibitor should be added to the glycol to prevent corrosion. Check for corrosion annually as glycol will deteriorate over a period of time. Flush and renew glycol if necessary.

If the existing hot water cylinder is heated by the central heating boiler then it will be necessary to install an additional pre-heat cylinder into the system. This is because in the winter months when the central heating is switched on the water in the cylinder will be automatically heated, and therefore will be hotter than the panels can heat it, resulting in no gain from the panels. If a pre-heat cylinder is installed, when hot water is drawn off the hot water cylinder it will be replaced by the solar heated water from the pre-heat cylinder, thus cutting down the amount of energy needed from the boiler. If the

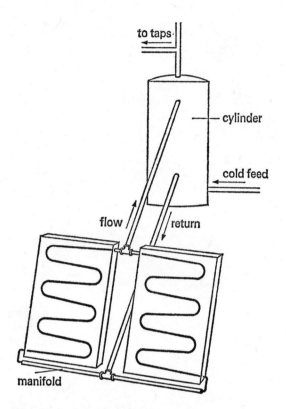

FIG. 33 *Typical thermo-syphon system using two panels.*

existing hot water cylinder has an immersion heater, then the panels can be connected straight to it. In the summer months most hot water will be supplied without using the immersion heater but in the winter months it will be needed quite often to boost the water temperature.

In a thermo-syphon system 22mm or 28mm ($\frac{3}{4}$″ or 1″) copper pipe should be used to connect the panels to the tank or cylinder. Larger diameter pipework will produce a better water flow rate to the storage vessel. Connect two compression reducers to the panel(s) inlet and outlet and connect pipes to them. Run the pipework at an angle up to the tank or cylinder as shown in Figs. 32, 33, 34.

Connect the pipework to the storage tank as follows. First drain off the water from the tank by opening the hot water taps in the system and closing the valve on the cold supply to the tank. Alternatively the taps can be opened and the ball valve in the tank tied in position so that it will not sink as the water level in the tank goes down. The

Pipework

ball valve will not open and fresh water will not enter the tank. Mark the positions for the two holes on the side of the tank. Next drill the appropriate sized holes in the side of the tank. Remove the large nut from the compression flange and insert the flange in the hole from inside the tank. Place the washer over the flange on the outside of the tank and then replace the nut onto the flange and tighten into place. The pipework can now be connected to the flange by an ordinary compression piece. The tank should be refilled slowly and the joints tested for leaks.

To connect the pipework to a cylinder, follow the same procedure, but drain the cylinder and the cold water storage tank that feeds it. Special flange connectors will be needed to ensure a watertight fit on the curved wall

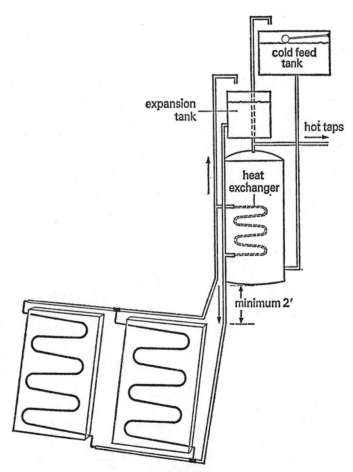

FIG. 34 *Typical indirect thermo-syphon system.*

of the cylinder. An alternative is to use the ordinary flange connectors as described above and solder the flange nut to the side of the cylinder. N.B. When fitting a flange connector to a cylinder the flange has to be inserted from outside the cylinder. This is done by drilling the appropriate sized hole and then distorting the sides of the hole to create a wider aperture for inserting the flange.

Indirect system heat exchanger

If an indirect system is used then a heat exchanger will have to be inserted in the cylinder. This should be done while the cylinder is drained. It should be positioned in the lower half of the cylinder and should not interfere with the immersion heater (the immersion heater can be temporarily removed if necessary). Basically a copper coil of approximately 1″ diameter tube, the heat exchanger is inserted into the cylinder by winding it through a pre-drilled hole. When the coil is completely inserted into the cylinder the two ends will emerge from the side of the cylinder and the pipework can be connected to it in the normal way. Yorkshire Imperial Metals Ltd., PO Box 166, Leeds, England, make the 'Sidewinder' heat exchanger coil which comes complete with template for drilling the holes in the cylinder, fibre washers, nuts and comprehensive instructions. Connections to the 'Sidewinder' can be made with compression or capillary fittings and the ends are threaded $\frac{3}{4}$″.

Make sure there are no leaks in the system before filling with water and anti-freeze. *Anti-freeze is poisonous and must not escape into the domestic water.*

INSTALLING A FORCED CIRCULATION SYSTEM

As in the thermo-syphon system, the panels can be connected to a pre-heat tank or cylinder or to the existing hot water cylinder. The same considerations regarding the use of the central heating system apply. It is advisable to use 15mm ($\frac{1}{2}$″) copper pipework through the system as it can be bent into desired shapes by using an ordinary spring bender. If your existing plumbing is iron then suitable copper to iron compression fittings must be used.

For most systems the pump can be a small central heating pump such as the SMC Commodore Range, Grundfos

Super 4/Super DW, Euramo, or Plessey. A slow flow rate is necessary for the panels to operate efficiently, so heavy duty pumps are not required. Likewise there is little pipework involved so the water resistance in the system will be quite low. With direct systems a *bronze pump must be fitted*. With indirect systems the cheaper standard pumps can be used as little corrosion is likely to occur. Small valves fitted either side of the pump are useful for restricting the flow rate and for isolating the pump if inspection or repair is necessary, A flow of 2 g.p.m. per panel is about right.

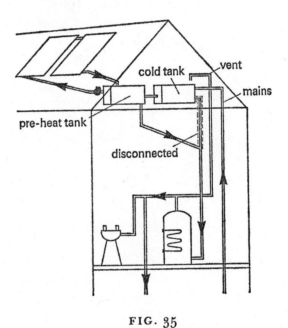

FIG. 35

DIRECT SYSTEM NO. 1 (FIG. 35)

Perhaps the easiest installation of all. Connect the pipework going to and from the panels to the existing cold water storage tank in the loft. As long as the cold taps in the household are fed by the rising main then the cold water storage tank will act as a pre-heat tank. Short runs of pipe in the loft space are all that is needed and a small pump (40 to 50 watts) can be used to circulate the water. This method has obvious advantages; no elaborate plumbing, no extra storage tanks or cylinders and the cylinder can be heated by auxillary means if required, having been automatically topped up with pre-heated solar water. Also, if the return pipe from the panels is kept just above the water level in the tank, then no automatic air vent is required to allow air into the system to facilitate draining when the pump cuts out. Air will enter through this pipe to facilitate draining.

There are, however, two disadvantages. First and most important, many local water authorities do not approve! Hot water could enter the mains supply if the ball valve in the tank failed to operate properly. Second, if the water in the cylinder is not heated and used for a period of time then many gallons will have to be drained off before the solar heated water enters the cylinder. Not much can be done about the second point, but the first problem can be overcome by fitting a separate cold water tank alongside the existing one and using it as a pre-heat tank to supply the cylinder. (Fig. 36). Use a galvanised or reinforced glass fibre tank similar in size to the existing cold water tank.

Cold water tank

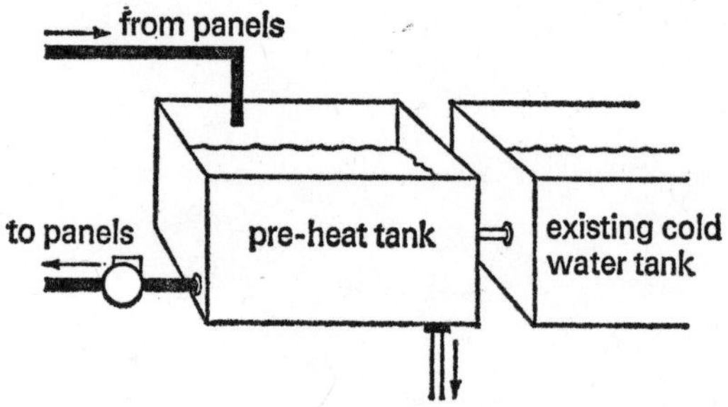

FIG. 36

It can be placed alongside the cold water tank and connected by a short run of 22mm or 28mm ($\frac{3}{4}$″ or 1″) copper pipe. A one-way valve should be inserted in the pipe to prevent reverse flow of hot water. Two flange tank connectors will be needed, and they should be positioned about 2″ from the bottom of the two tanks. (Don't forget to drain the tank first!). The feed to the hot water cylinder from the cold tank should be disconnected and transferred to the new solar tank. See Fig. 35. The new pre-heat tank is now ready for the pipework from the solar panels on the roof. Fit another flanged tank connector to the tank 4″ (100mm) from the bottom and using 22mm (¾″) copper pipe connect to the pump which should be supported on a bracket or small block of wood. The other end of the pump can then be connected to the flow pipe to the panel. The return pipe from the panels should enter the top of the tank above water level. This will allow air to enter the system when it is being drained. The panels must be above the storage and pre-heat tanks to allow for draining in cold weather, and all pipework must be lagged to prevent freezing.

Minimum heat loss

DIRECT SYSTEM NO. 2 (FIG. 37)

Another easy system to install. The panels are connected directly to the existing hot water cylinder. The main advantage of this method is that the solar heated water is used direct from the cylinder and heat loss is therefore kept to a minimum. However, if central heating is in

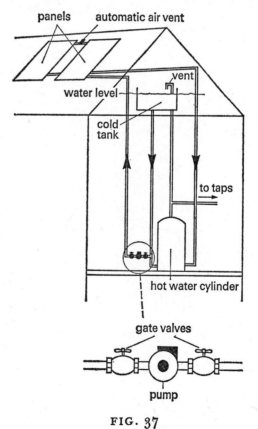

FIG. 37

operation, the water in the cylinder will be warmed anyway and no benefit will be received from the panels. If the cylinder is heated by an immersion heater then this can be turned on occasionally to boost the solar heated water when it has not reached the desired temperature.

The first step is to bring the pipework down from the panels to the cylinder. There will usually be a hole in the ceiling where the existing cold water feed comes through to the cylinder. This hole can be enlarged to accommodate the new pipework. Next drain the cylinder as described earlier in the chapter. A hole must be drilled in the cylinder wall about 6″ from the bottom and opposite the cold

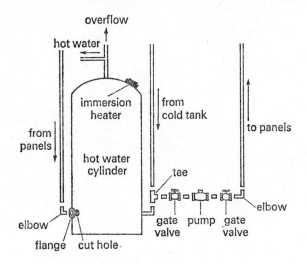

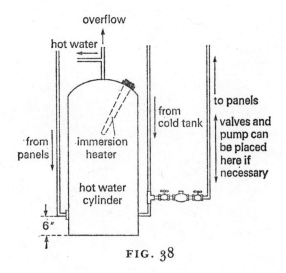

FIG. 38

water inlet side if possible (Fig. 38). A flanged connector to take the 15mm ($\frac{1}{2}$") pipe from the panels should then be fitted and the pipework connected to it. Next cut the cold water inlet feed roughly 50mm (2") from the elbow and remove about 35mm ($1\frac{1}{2}$") of pipe. A compression tee fitting should then be inserted in the gap with the 15mm ($\frac{1}{2}$") end at right angles to the pipe. It is to this connection that the flow to the panels is joined, and the pump (with valves) is connected to this section either close to the cylinder, or up in the loft (the pump must be situated below the cold water level in the tank). An automatic air vent must be fitted to the panels at the highest point; i.e. between the two panels in a two panel system, or on the top header pipe in a multi panel system. The vent will allow air to enter when the system is draining; otherwise air locks might occur resulting in possible damage.

In this type of installation the water will drain down to the level of the cold water tank so it is essential that all pipework in the loft be insulated. Small amounts of water may escape through the overflow when the system is draining but this is not harmful and should present no problems.

INDIRECT SYSTEM NO. 1

Heat exchanger

This entails connecting the panels to the existing hot water cylinder. The cylinder must have enough room inside it to accommodate a 'Sidewinder'. Alternatively it may already have a heat exchanger fitted to which the panels can be connected. If the cylinder already has a heat exchanger connected to the central heating boiler then it may not be possible to add another heat exchanger, and a separate pre-heat cylinder will be needed (Indirect system No. 2). Some indirect cylinders have the capacity to accommodate another heat exchanger, but advice should be obtained before attempting this.

Following the circuit in Fig. 39, bring the pipework from the panels to the heat exchanger in the cylinder (see passage on installing a heat exchanger p. 52). Make sure that the pump is on the flow to the panels and that the header tank is connected to the return pipe. This will ensure that the pump does not 'pump-over' through the expansion tank which would cause oxygen pick-up in the water and lead to corrosion in the system. An automatic air vent is not necessary in this system. However, it is advisable to fit a pressure relief valve (in the same position as an automatic air vent) to protect the system from boiling in very hot weather.

INDIRECT SYSTEM NO. 2

Separate pre-heat cylinder

If it is impracticable to connect the panels to the existing hot water cylinder, a separate pre-heat cylinder should be used. This additional cylinder can be placed alongside the existing cylinder or situated above it in the loft (Fig. 39). A cylinder of the same size and capacity as the existing cylinder should be used and should preferably have a heat exchanger already fitted. Disconnect the cold water supply to the existing cylinder and run it to the pre-heat cylinder. Next connect the hot water outlet from the top of the pre-heat cylinder to the inlet on the existing cylinder. Finally connect the pre-heat cylinder to the panels as described previously. The pre-heat cylinder will fill with cold water from the cold storage tank which will be heated by the panels through the heat exchanger. When water is drawn

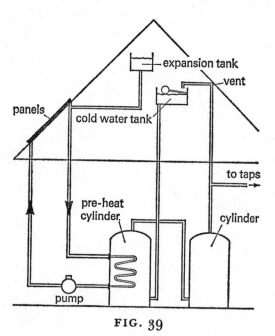

FIG. 39

from the existing cylinder it will be replaced with the solar heated water. The immersion or boiler can be left on to heat the water to the preferred temperature, but its work load will be cut as it will no longer be heating mains cold water. When the solar water is hotter than the setting on the immersion or boiler thermostat then it will not cut in at all and you will be relying 100% on solar heated water (Fig. 40).

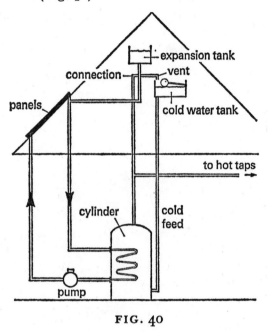

FIG. 40

Summary

Important considerations with indirect systems are:

(a) When filling the system it is advisable to use deionised or distilled water with a suitable anti-freeze such as ethylene glycol and a suitable inhibitor to prevent cupro-solvency. (Many anti-freeze solutions contain inhibitors but it is wise to check).

(b) Make sure that there are no leaks in the closed circuit. Anti-freeze is poisonous!

(c) Check the pipework and panels annually for signs of degrading due to acidity in the closed circuit fluid. If necessary flush out the system and re-fill with a fresh mixture of water and anti-freeze.

(d) Allow extra panel area to compensate for reduction in efficiency due to using heat exchanger.

4 Control Devices for Solar Equipment

Reverse flow

In thermo-syphon systems the heated water will flow naturally to the insulated storage tank or cylinder. Even when the panels become cooler than the storage tank, reverse flow will not occur as long as there is a minimum distance of 60 cm (2ft.) between the top of the panels and the bottom of the tank. A non-return valve can be fitted to prevent the downward flow of water. In cold climates with freezing temperatures adequate provision must be made for frost prevention. Shutters or covers for the panels become a nuisance and do not always give protection in very cold conditions. It is far better to use an indirect system (Fig. 34) and introduce anti-freeze (ethylene glycol) into the system via the separate expansion tank.

Frost prevention

In forced circulation systems it becomes necessary to have a method of turning the pump on and off as required. In very warm climates the pump can be switched on manually first thing in the morning and off late in the afternoon or early evening, or a simple time switch could be used. Times of sunrise and sunset can be obtained locally and the automatic time switch set accordingly. Ninety per cent of solar radiation is collected in the middle two thirds of the day.

Time switch

In slightly cooler climates a more sophisticated control is needed to match the changes in available radiation. A couple of sunny days could be followed by a day of cloud and rain, cooling the panels considerably. Once water has been heated it needs to be stored and not pumped back to cold panels where it will rapidly cool down. A simple

Typical Plumbing Fittings
1. *15mm Tank Flange—Compression*
2. *15mm × 15mm × 15mm Equal Tee—Compression*
3. *15mm × 15mm Straight Connector—Compression*
4. *15mm × 15mm Straight Connector—Capillary. Also shown 10mm × 10mm Straight Connector—Capillary*
5. *10mm × 15mm Reducer—Compression/Capillary*
6. *Mustang Air Vent, Balsa Type*
7. *Flexivent Automatic Air Vent, Positioned on 15mm Equal Tee (iron).*

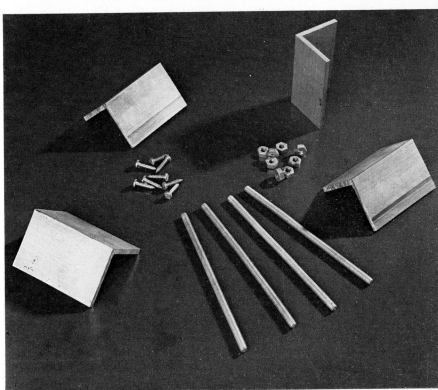

Four aluminium brackets with cadmium plated self tapping screws, screwed rod and nuts.

61

Measuring the roof and drilling holes to accept the bolts for attaching the solar panel.

Taking the panel up on to the roof.

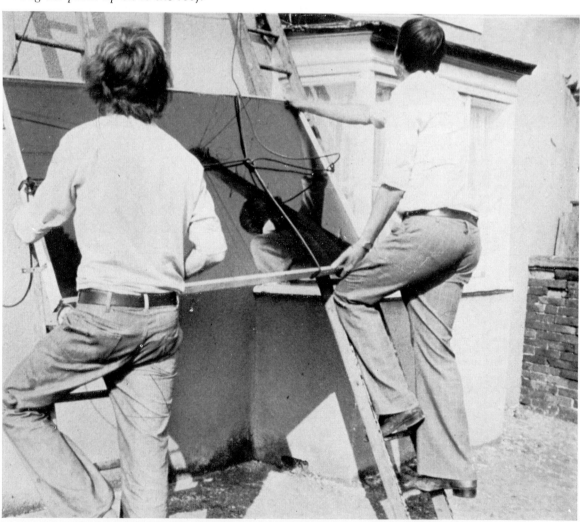

A typical differential temperature controller incorporating the circuit described in this book.

A typical circulating pump being positioned for a direct type solar system. This pump has integral valves.

Two typical circulating pumps. The one on the left is a bronze type for direct or 'open' systems of solar water heating, and the one on the right for use only on indirect or 'closed' systems.

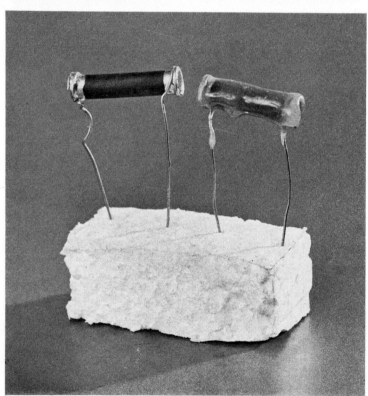

Insulating the thermistors. The left hand side thermistor is untreated, the other has been coated with Araldite epoxy resin.

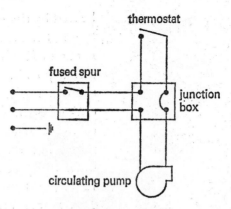

FIG. 41 *Wiring diagram for thermostat.*

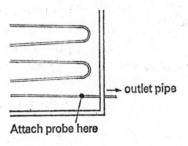

FIG. 42

Thermostat

thermostat can be wired to the pump and main (Fig. 41). It will cut out and switch the pump on at a pre-selected temperature e.g. 24°C (76°F). This means that when there is useful heat to be collected the pump will be switched on, and whenever the panel temperature falls below the given temperature the pump switches off preventing stored hot water from being cooled down too much.

The thermostat should be placed near the panel(s) in a weather-proofed area, and the thermostat probe attached to the outlet pipe inside the panel as in Fig. 42. A 3 amp fuse should be incorporated in the circuit.

While this method of control works fairly well in warm climates it does have a number of drawbacks in more northerly latitudes. Let us assume the system is working from scratch and the water in the storage tank or cylinder is cold, having come from the mains. If the water temperature is for instance 13°C (55°F) and the panels have been heated by the sun to produce a water temperature of 21 °C (70 °F) then the pump will not operate as the temperature needed to activate it (24 °C) has not been reached. Therefore a possible heat rise of 7 °C (20 °F) has been lost. On the other hand, if the water temperature in the storage tank or cylinder has reached 54 °C (130 °F) and the panels then begin to cool rapidly, the pump will continue to pump hot water to the panels, where it will be cooled, until the pump cuts out at 24 °C (76 °F). A heat loss of 12 °C (54 °F) will have occurred and it might be several days before enough radiation is available to boost the temperature again. Therefore in northern latitudes a more critical control device is needed; one that will operate the circulating pump whenever there is a small increase in temperature to be gained and cut out the pump whenever the panels cool down.

Differential temperature controller

That device is called a differential temperature controller. It consists basically of an electronic circuit with four main terminals. The first terminal takes a mains input;

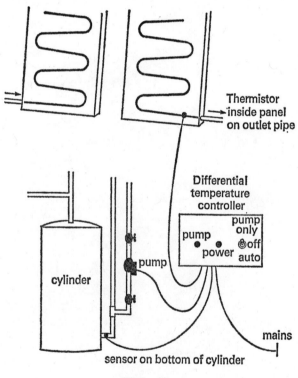

FIG. 43

COMPONENTS FOR A DIFFERENTIAL TEMPERATURE CONTROLLER

R1   2.7k ohm   0.5W Carbon Resistor
R2   2.7k ohm   ,,   ,,   ,,
R3   1.0k ohm   ,,   ,,   ,,
R4   1.0k ohm   ,,   ,,   ,,
R5   3.3k ohm   ,,   ,,   ,,
R6   100k ohm   ,,   ,,   ,,
R7   68k ohm   ,,   ,,   ,,
R8   33k ohm   ,,   ,,   ,,
R9   3.3k ohm   ,,   ,,   ,,
RL   Omron 8-pin Plug-in Relay 12V DC nom.
IC   Integrated Circuit 8-pin 741
Z1   Zenner Diode 10V
Z2   Zenner Diode 3.3V
TR1 PNP Transistor
VR1 Cermet Trimmer RS Type G. 1.0k ohm.

also required

Three-way Switch
Two Neons
Vero Board
Enclosure
12-position Terminal Block
3-core Wire
2-core Screened Wire

Thermistors should be types VA 1005 or TH 2a
and should be matched at ambient temperature.
Thermistors type VA 1050 can also be used.

the second takes the mains current to the circulating pump. Between these two terminals is a relay switch that turns the current on and off. It is controlled, through the circuit, by the other two terminals which sense changes in temperature. The third terminal is connected to a thermistor situated in the panel; the fourth is connected to a thermistor situated in or at the bottom of the storage vessel (Fig. 43). Whenever the thermistor in the panel is warmer than the thermistor in the storage tank or cylinder, the relay will be closed sending the mains current to the pump which in turn sends the cooler water to the panel to be heated. Whenever the thermistor in the panel is cooler than the thermistor in the storage vessel the relay opens cutting off the electric supply to the pump.

There are now many firms manufacturing differential temperature controllers. Most of these firms are in the U.K. or the United States, but the author has received details of control boxes manufactured in Australia, Japan, and South Africa. They are not usually cheap; the average price in the U.K. is about £35 and in the United States about $80. However, with mass production techniques these figures might be halved within the next few years. If purchasing a commercial control box, shop around! Compare prices. Does the controller have neons to indicate that the pump is switched on? Does it have a plug-in relay for easy servicing? An on/off switch? Is it guaranteed?

Commercial control boxes

Making a differential temperature controller

For those of you who can read a circuit diagram and use a soldering iron I have included details of a typical differential temperature controller (Fig. 44) which you

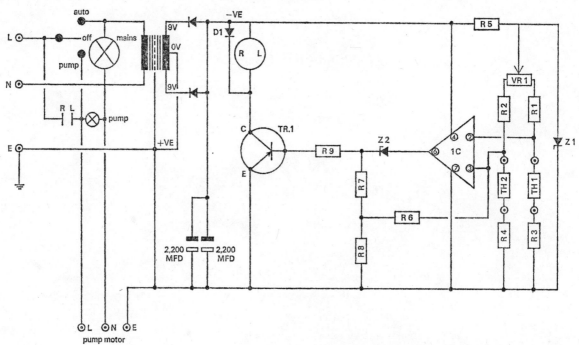

FIG. 44 *Temperature differential control circuit*

68

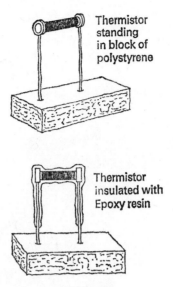

Thermistor standing in block of polystyrene

Thermistor insulated with Epoxy resin

FIG. 45

can make for around £12.00 or $20.00. This is only for those who really know electricity. YOU HAVE BEEN WARNED! The circuit can be mounted on Vero board and housed in a small plastic or aluminium box. Two neons can be attached, one to indicate that power is connected to the box, the other to indicate that the pump is working. The switch has three positions. The top position (PUMP ONLY) overrides the controller and provides a permanent current to the pump. The middle position (OFF) isolates the controller from the mains, and the bottom position (AUTOMATIC) operates the differential.

Wiring thermistors

When wiring the thermistors, use screened twin-core cable. Attach equal lengths of cable to each thermistor. Unequal lengths can cause changes in resistance resulting in faulty temperature readings. As the thermistors will be carrying a small voltage they must be insulated. Otherwise when they touch metal or water a short circuit will occur. The best method of insulation is to treat the two thermistors with a good epoxy resin such as Araldite. The thermistors wires can be stood in plasticine or a small block of polystyrene (Fig. 45) and the epoxy resin applied to the rod and surrounds. Make sure that the whole thermister is covered. Insulation tape, plastic tube, and small plastic bags could also be used to insulate the sensors. The thermistor situated in the panel should be attached to the outlet pipe as in Fig. 43. This will sense the hottest water temperature in the panel. The other thermistor should be situated at the bottom of the storage tank or cylinder to sense the coldest temperature.

A potentiometer in the controller can be adjusted to widen or close the differential between the two thermistors. A very small differential of say 1°C (2°F) may result in the pump 'hunting' throughout the day and night resulting in an unsatisfactory performance. A very wide differential will result in reduced efficiency of the system. The author recommends a set differential of about 2°C (5°F).

A good method of testing the differential is to immerse the insulated thermistors in two respective cups of hot water, one cup being slightly hotter than the other and the temperatures recorded by two thermometers. Alter the potentiometer until the relay switches at a 2°C (5°F) differential. The thermistor that switches the relay (this is indicated by the 'pump' neon) and turns the circulating pump on is the one to place in the panel.

It is important that the control box be protected by a 1 amp fuse, otherwise an electrical fault in the box could result in fire damage.

Circulating pumps up to 250 watts can be operated by the controller. Most domestic pumps will have ratings of between 40 and 100 watts. Heavy duty pumps will need further electronic equipment in addition to the controller and expert advice should be sought.

All the components are obtainable from amateur radio shops, mail order companies specialising in the supply of electronic components, or from the local electrical shop.

A differential temperature controller similar to the one described is obtainable from Maybeck Ltd., Foster House, Studley, Warwickshire, who specialise in the supply of D.I.Y. solar equipment.

5 Swimming Pool Heating

Heating a swimming pool with solar energy is an attractive proposition as most pools are situated outdoors and are therefore used in the summer when the sun is hotest. The economics of solar-heated swimming pools are very attractive when compared to the cost of conventional fuel heating. Although there is a much greater volume of water to be heated than in a domestic hot water system, it is not necessary to use a correspondingly greater number of panels because the water temperature does not need

to be so high. In fact it is probably true to say that the majority of solar water heating systems in operation throughout the world heat swimming pools. In some areas of the United States it is illegal to use gas to heat pools and this has resulted in a boom for the local solar heating companies.

Swimming pool ownership is now quite common in the United Kingdom but heating a pool costs money. If you want the water temperature to be higher than nature intends it to be, then you have to supply continuous heat to counteract the loss caused by evaporation and convection. This results in heavy bills when using conventional fuels such as solid fuel, gas or oil. And there is the expensive equipment needed to supply heat to the pool.

If your swimming pool already has conventional heating then you can add the solar heating system to it and make use of the existing pipework, pump and filter, thus reducing your outlay. If your swimming pool has no existing heating then you will save yourself the cost of an expensive boiler installation. Either way you will be using the sun's free energy instead of expensive fuels that will probably more than double in price in the next ten years.

The panel described in this manual is ideal for swimming pool heating. Having copper waterways, it will not be affected by chemicals in the water and other impurities associated with pools. It is also self-draining and can therefore be mounted above the pool water level to facilitate draining in cold weather. Panels can be mounted on the roof of your house or an outbuilding, or on a specially constructed frame.

HOW MANY PANELS?

There are two common methods of selecting the number of panels to be used; (1) relating to area of the pool, (2) relating to the volume of the pool. Norman Sheridan, in his paper 'Solar Heating a Swimming Pool', *Solar Research Notes No. 6* (see appendix), recommends a collector area equal to the area of the pool surface. Francis de Winter, in his booklet *How to Design and Build a Swimming Pool Heater* (see appendix), recommends a panel area equal to one half of the pool's surface area. Many manufacturers equate their panel area to the gallonage of the pool and their recommendations vary widely. Some solar panels designed for swimming pool use are not glazed and therefore need a larger collection area. A more efficient panel design will need less area. If too little area is used then evaporation and convection losses prevent the attainment of reasonable temperatures. If too much area is used then the panels will overheat and lose energy to the surrounding atmosphere, resulting in a diminished return on investment. Panel area must therefore be decided on the basis of pool area or volume in relation to panel efficiency.

Calculation

So what is the correct method to calculate the numbers of panels needed? The author recommends the following procedure. First take the area of the pool surface (e.g. 300 sq. ft.) and divide by two (150 sq. ft.) = 10 panels. Next work out the gallonage of the pool by multiplying the area by the average depth, e.g. 300 × 4ft. 6in. = 1350 cu. ft. Multiply this answer by 6.23 to find the number of gallons; e.g. 1350 × 6.23 = 8411 gallons. Next divide the gallonage by 1250 (i.e. one 15 sq. ft. panel per 1250 gallons) and you get the answer 7. Take whichever number is the greater e.g. 10, and that is the number of panels to use.

For swimming pools in the British Isles or similar latitudes (50° to 60° N) solar panels are best glazed to give maximum efficiency. In warmer climates panels need not be glazed as there is more sunshine through the year and ambient air temperatures are much higher.

INSTALLATION

Selection of site

First select your site. If possible mount the panels on an outbuilding or nearby roof. They will be out of the way of swimmers and the cost of a special framework is avoided. Mount the panels so that they do not obstruct sunshine falling on the pool, which also helps raise the water temperature (Fig. 46). As the panels will be heating the pool mainly in the summer months it is advisable to use a lower angle of inclination than in a domestic system where energy collection is required all the year round.

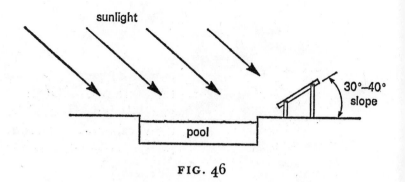

sunlight

30°–40° slope

pool

FIG. 46

Angle of latitude less 10° or 15° will give slightly better efficiency in the summer period. If mounting the panels on a roof then you will be governed by the slope of the roof. If necessary the panel mounting brackets can be packed with neoprene gasket or battens to achieve small variations in angle.

Making a frame

If a frame is needed this can be made from a good hardwood 2″ × 3″ hammered into the ground as shown in Fig. 47. Make sure the posts are very secure! Frame mounted panels catch the wind, and in severe gales they might be damaged if not secured properly. Use large wood

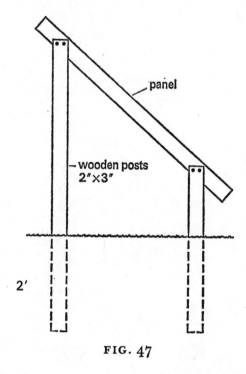

panel

wooden posts 2″×3″

2′

FIG. 47

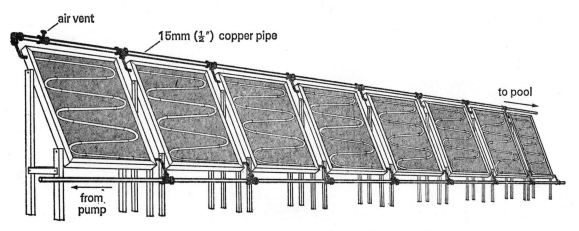

FIG. 48 *Panels mounted on frame.*

screws, two per post connection. Don't forget to treat the wooden frame with protective coating. Aluminium angle, Dexion angle and steel tube can also be used to make a suitable frame.

Connecting panels

Panels should be connected in parallel (Fig. 48), the 10mm ($\frac{3}{8}$″) inlets and outlets being connected to 15mm ($\frac{1}{2}$″) copper manifolds as shown. 10mm × 15mm reducers and a 15 mm elbow could connect to a 15mm equal tee on the manifold. Alternatively, the 10mm inlets and outlets could be made longer (about 7 or 8 inches) and carefully hand bent to connect to a 15mm ($\frac{1}{2}$″) × 15mm ($\frac{1}{2}$″) × 10mm ($\frac{3}{8}$″) tee on the manifold.

It is unwise to connect many panels in series. In very hot weather the water could be boiling by the time it reaches the last panel and this might damage the panels. Also panels linked in series require a slower flow rate compared to panels linked in parallel, thus increasing the time for the pool water to circulate and reducing efficiency.

Drainage

It is not necessary to lag the pipework, unless long pipe runs are involved, as the system will be used in warm weather when heat loss is low. However, the system must be able to drain in cold weather conditions to prevent water freezing in the panels. If you position the panel array above the level of the pool, water will automatically drain from the panels and into the bottom 15mm ($\frac{1}{2}$″) manifold. From there it will return to the pool or be re-

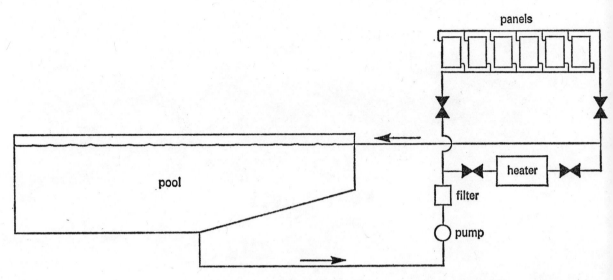

FIG. 49 *Plumbing diagram involving solar panels and existing heater.*

leased through a drain-valve. The bottom manifold should be on a slight incline to help water drain. An automatic air vent will be necessary to allow air into the panels to assist drainage. It should be placed on the top manifold at the position shown in Fig. 48.

CONNECTING TO EXISTING HEATING SYSTEMS

Fig. 49 shows a typical plumbing diagram involving solar panels and an existing heater. The various valves can be manipulated to direct flow as required; through the existing heater, through both heater and solar panels, or through solar panels only. The heater can therefore operate independently when there is little solar radiation, or it can be shut down completely when there is enough radiation for the panels to heat the pool by themselves.

CONNECTING TO EXISTING FILTER SYSTEM

Figs. 50 and 51 show typical plumbing diagrams involving solar panels and existing 'filter only' systems. In Fig. 50 a separate pump is shown. This is sometimes necessary if the existing pump is too powerful or the filter will not take the slower flow rate needed for efficient use of the panels. When the solar pump cuts out in cold weather the water will continue to circulate through the filter thus staying fresh.

In all systems it is important to be sure that the panels can be drained in very cold weather, to avoid the possibility of water freezing in the pipework.

Make sure that circulating pump is BRONZE otherwise corrosion will occur.

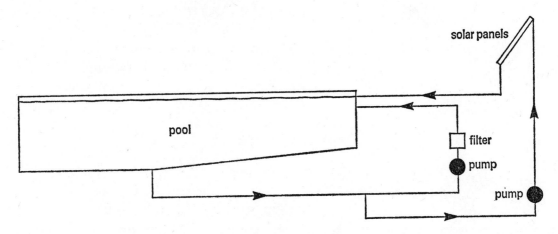

FIG. 50 *Plumbing diagram involving solar panels and existing filter system.*

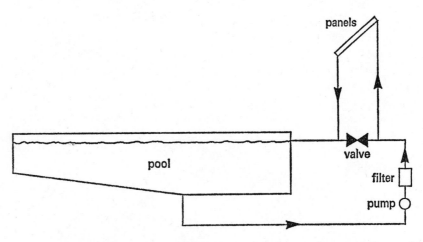

FIG. 51 *Plumbing diagram involving solar panels and existing system.*

FLOW RATES

The flow rate of water through the solar panels is not critical, but very high flow rates tend to be as inefficient as very slow rates. Water passing through the panels too quickly will lower the panel temperature indicating that heat is being extracted, but at the same time evaporation losses are increased from the pool. A very slow flow rate will result in high water temperatures from the panels, but the pool volume will not be circulated quickly enough to make the increase felt. The pool volume should be circulated every eight hours and this usually means a flow rate through each panel of between 2 and 3 gallons per minute. A 10 panel installation connected in series would therefore have a flow rate of between 20 and 30 gallons per minute.

CONTROL OF CIRCULATION PUMP

This can be achieved by manual operation, simple thermostat, time switch, or differential controller.

Manual operation is often used in solar pool heating systems for unlike domestic water heating systems slight changes in temperature can be tolerated without much adverse effect on the overall daily gain. Thermostats are also widely used even though they cut in or out at a single fixed temperature. Time switches have been used to start the circulating pump in the morning and shut it off in the late afternoon. A differential controller will provide the best results as it will automatically sense energy rises and decreases in the panels and operate the pump accordingly. The controller described in Chapter 4 can be used and will control pumps up to 250 watts. One sensor should be placed in the pool; the other on the outlet pipe inside one of the panels. Remember that the filter may still need to operate when the solar system is not in use.

POOL COVERS

If your pool can be covered then heat loss will be reduced. A transparent cover will allow solar energy to enter the pool helping to raise the temperature. It will also prevent heat loss during the day and night. If your pool is indoors or has a transparent dome cover then heat loss will be kept to a minimum. However, if your pool is out of doors you will need an easily removable cover, otherwise heat gained during the day will be lost at night and the solar panels will have to start the previous day's cycle all over again.

A properly installed solar heating system for your pool will save you money on heating bills and will extend your swimming season by about two months. Temperatures between 21° and 27°C (70°–80°F) can be achieved during the summer in the British Isles by using solar panels. In warmer climates even higher temperatures can be obtained. It will also take some of the worry out of pool ownership and will protect your investment. After all, your pool is meant to be enjoyed as a means of relaxation and creation, not regarded as a financial burden.

6 Conclusions

The success of your solar heating system will be measured by the money you save on fuel bills.

The author knows of several people who have built systems from the plans in this book (the plans were originally drawn in 1975), and they have had varying degrees of success. A man who installed two panels at his home in Malta now never uses his gas water heating, but relies on solar heated water alone for all his domestic hot water requirements. Another two-panel installation in northern Scotland has provided 100% of the hot water requirement on only a few days of the year in 1976. Two and three-panel installations in the south of England seem to be working very well and generally fall within the predictions made in Chapter 1.

For those of you who would like a solar heated system but do not wish to build and install the panels yourself, the following advice may prove helpful. There are still no standards or manufacturing codes applicable to solar equipment. There are no British Standards and in the United States, Congress is still considering proposed standards. Before purchasing commercially available equipment you should satisfy yourself that the claims made by the manufacturers are correct. Flat plate collectors should have a reasonable efficiency in relation to cost. Any supplier who cannot give an efficiency rating for his equipment should be avoided. Either he hasn't tested his panel design or he is too ashamed to give you the results. Satisfy yourself that the materials used in a panel will withstand the high temperatures it may reach during very sunny weather, and that it will not degrade rapidly. Generally speaking, panels with aluminium or copper absorber plates last longer than other types. Copper waterways are considered best and will give few problems. Aluminium, steel, and plastic radiator panels do not stand up very well as a rule and usually degrade quite rapidly after ten to twelve years. This is because they are affected by oxygen in the water which causes corrosion, or by the acidity in anti-freeze solutions. The panel to consider is one that has copper waterways, is able to withstand internal temperatures of 204°C (400°F) for periods of six hours at a time, and is constructed of materials that will not degrade by more than 5% during a ten year period. It should also have a selective black coating.

Installation firms offering a package deal are becoming more numerous and while many of them are honest and know what they are doing there are bound to be the few, as in all trades, who will charge excessively and provide an inferior installation. If you decide to place an order with an installation company try to find out how long they have been established in the solar business (reputable installers will usually belong to a trade association or will be members of the International Solar Energy Society), and don't part with all of your cash until the job has been completed properly.

Solar energy is going to play an increasingly important role in all our lives. Perhaps this book will provide some impetus to the movement, and hopefully some substantial savings in YOUR future fuel bills.

Best of Luck.

Appendices

Appendix 1
Further Reading

BOOKS

Brinkworth, B. J., *Solar Energy for Man*, Compton Press, Tisbury, England, 1972. ISBN 9001-9313-1.

Daniels, Farrington, *Direct Use of the Sun's Energy*, Yale University Press, 1964. ISBN 345-23794-3-195.

De Winter, Francis, *How to Design and Build a Solar Swimming Pool Heater*, Copper Development Association, New York, 1976.

Duffie, John A. and Beckman, William A., *Solar Energy Thermal Processes*, John Wiley, New York, 1974. ISBN 0-471-22371-9.

Halacy, Jnr., D. S., *The Coming Age of Solar Energy*, Harper and Row, New York, 1964.

Halacy, Jnr., D. S., *Solar Science Projects*, Scholastic Book Services, New York, 1974.

A Question of Solar Heating, Copper Development Association, Potters Bar, England, 1976.

Sheridan, Norman R., *Solar Heating a Swimming Pool, Solar Research Notes No. 6*, University of Queensland, Australia, 1975.

Solar Energy – A U.K. Assessment, International Solar Energy Society, U.K. Section, London, 1976. ISBN 0-904963-08X.

Williams, Richard J., *Solar Engery, Technology and Applications*, Ann Arbor Science, Michigan, 1974. ISBN 0-250-40142-8.

Alternative Sources of Energy. Box. 36 – B. Minong, W15 54859, U.S.A.

Development of Proposed Standards For Testing Solar Collectors and Thermal Storage Devices. (National Bureau of Standards – Washington). No. 899, 1976. U.S. Government Printing Office, Washington D.C. 20550.

Solar Energy. A quarterly Magazine published by Pergamon Press for the International Solar Energy Society.

The following are available from Brace Research Institute, MacDonald College, McGill University, Quebec, Canada.

> *How to Make a Solar Still.*
> *How to Make a Solar Steam Cooker.*
> *How to Make a Solar Cabinet Dryer – Agricultural.*

USEFUL ADDRESSES

Arden Publishing, 2 Redditch Road, Studley, Warwickshire. Import and stock books on Solar Energy and specialise in mail sales of same.

Geliotekhnika. Bi-monthly Magazine of Academy of Sciences of the Uzbeck SSR. (English translation – Applied Solar Energy by Allerton Press).

International Solar Energy Society. Headquarters: PO Box 52 Parkville, Victoria, Australia 3052.

International Solar Energy Society – UK Section. c/o Royal Institution, 21 Albermarle Street, London, W1X 4BS.

National Centre for Alternative Technology, Macynlleth, Wales.

Appendix 2
Conversion Tables

WATER CONTENT OF COPPER TUBE

Copper Tube		
Size	Water Content	
	Gal/m	Litre/m
6mm	0.004	0.018
8mm	0.008	0.036
10mm	0.013	0.059
15mm	0.032	0.146
22mm	0.070	0.318
28mm	0.118	0.837

CONVERSION FACTORS

To make the required conversion × or ÷ by appropriate factor.

Areas

Sq. in. to Sq. cm.	Sq. cm. to Sq. in.	Sq. ft. to Sq. m.	Sq. m. to Sq. ft.	Sq. yd. to Sq. m.	Sq. m. to Sq. yd.
6.452	0.1550	0.0929	10.76	0.8361	1.196

Volumes

Liquid Ounces to Cu. cm.	Cu. cm. to Liquid Ounces	Pints to Litres	Litres to Pints	Gallons to Litres	Litres to Gallons
29.57	0.03381	0.568	1.66	4.546	0.22

Power

Horsepower to Kilowatts	Kilowatts to Horsepower	Metric Horsepower to Kilowatts	Kilowatts to Metric Horsepower
0.7457	1.341	0.7354	1.360

Horsepower to Metric Horsepower	Metric Horsepower to Horsepower	Megajoules to Kilowatts
1.014	0.9863	3.6